DISABILITY AWARENESS
© Copyright 2015 DANIELE LUCIANO MOSKAL

ISBN: 978-1-326-38309-1

Daniele Luciano Moskal asserts the moral right to be identified as the author of this work. All rights reserved.

[First Edition]

Printed in the United States of America

Email: penofareadywriter@hotmail.co.uk
Website: www.dlm-evangelical-ministries.org

All rights reserved under International Copyright Law. No part of this book or its contents or its cover may be reproduced or transmitted in whole or in part in any mechanical or electronic form including photocopying, recording, or by any other information or retrieval system without the express written consent of the Author and Publisher of this first edition book by the 'penofareadywriter' - *Daniel Luciano Moskal.*

CONTENTS

Page 1…………………………………Introduction
Page 3……………………………….Note of thanks
Page 4……………………………..Visual Disability
Page 21……………………………Hearing Disability
Page 24………………………The Deaf and the Blind
Page 26…………………………… Multiple Sclerosis
Page 28……………………….Rheumatoid Arthritis
Page 29……………………………Alzheimer Disease
Page 32…………………………Learning Disabilities
Page 34……………………………...Mental Illness
Page 37……………………………...Schizophrenia
Page 41………………………………………..Strokes
Page 48………………………..Parkinson's Disease
Page 49…………………………………………Epilepsy
Page 53…………………………………….Paraplegia
Page 54……………………………………….Old Age
Page 59……………………….Muscular Dystrophy
Page 62……………………………….Osteoarthritis
Page 63………………….Calling for an Ambulance
Page 64………………...About The Author/Thanks

INTRODUCTION

What does the word 'Disability' mean?

In writing this booklet, I hope and pray you the reader will get a better understanding by simply looking at some of the disability categories which you may experience when coming across people with disabilities, and hopefully consider how best you can respond to their needs should the moment arise. Firstly, referring to the "disabled" just does not help. We are talking about real people first and foremost – people who are daily presented with extra challenges because of a range of impairments they are either born with, or sustained through an accident, particular circumstances, illness or old age. Many people with a disability may not have access to cars and will therefore rely heavily on the usage of public transport to provide mobility and independence. It is vitally important that you and I the reader understands and appreciates the needs of people with disabilities. The effects of some impairment will be immediately obvious to you, but others may be invisible to the naked eye or be easily mistaken for something else. Let us now take a look at some of the facts and disabilities in today's society.......

NOTE OF THANKS

I would like to say a very big thank you to **'MENCAP'** the leading voice of learning disability in the United Kingdom. Everything they do is about valuing and supporting people with a learning disability, and their families and carers. Having worked for them voluntarily in the late 80s twice as part of their Summer Play Scheme –

I had the privilege of working with two young teenagers who both had learning difficulties, and were unfortunately both prone to violent mood swings; blackouts and epileptic fits. I learned a great deal and became more aware about disability. It was a most humbling and worthwhile rewarding experience that I recommend anyone regardless of age should participate in. Thank you MENCAP! This is their website address: **https://www.mencap.org.uk/**

VISUAL DISABILITY

Visual impairment can range from various forms of poor sight to complete blindness. Some, but not all, visually impaired people will carry a white cane (stick), to aid them, while others may have a guide dog. Visual impairment people need to be helped; especially when they need to catch a bus. So this is were you can be of valuable assistance by waving down a particular bus, and calling out the destination of the bus as they board, and your voice will also help to give them a sense of direction.

It is a good idea if you are boarding the same bus to check the customer's destination stop as they enter, and also inform the driver when they are approaching their stop. Remember you can see if a seat is available – a blind person cannot! It does not take much effort to advise a visual impairment person exactly where a seat is available, bearing in mind the person may have a guide dog. It goes without saying that the bus driver should wait a few seconds until the visual disabled person is comfortably seated before they move off!

Any approaches to a blind person (from you or any other person(s)) should always be made to the side opposite the guide dog. Guide dogs all though cute and caring should not be petted or stroked, as they could get distracted with serious consequences. Once the destination stop has been reached it helps if you can give the visual disabled person an indication of how far the bus is from the kerbside.

When the bus pulls up, make sure when disembarking the bus that there are no hazards like lampposts, stop signs, or railings the person could easily collide with it. Again, if you are the one helping the visual impaired person you should wait until he/she is safely on the pavement and that there are no other dangers from bus exhaust fumes, or even the visual disabled person stepping back into the vehicle before leaving them to go about their business.

What is blindness? **"Blind"** means a high degree of vision loss, for example seeing much less than is normal or perhaps nothing at all. All though the word *'blind'* is widely used to imply total vision loss, only about 18% of people registrable as blind have no useful sight. The remaining 82% have some useful, residual vision.

Some people who are registered as *'blind'* may be able to see an object three metres away. A person is registrable as blind, according to the National Assistance Act (1948) if they are considered to be *'so blind as to be unable to perform any work for which sight is essential'*. This does not refer exclusively to employment, but to any activity for which sight is required.

What is partial sight? The term 'partial sight' also refers to sever loss of vision but to a lesser degree than blindness. It generally means not seeing very well but having enough sight to read and write. There is no statutory definition for partial sight under the 1948 Act but the Ministry of Health advises that to be registrable as partially sighted a person must be substantially and permanently impaired by their defective vision.

What is impaired vision? Impaired means that something does not work as well as it might. People with impaired cannot see very well. People who have severely impaired vision are therefore registrable as blind or partially sighted. Some people do not like to be labelled as *'blind'* or *'partially sighted'*. They feel that these terms are negative and misleading. For this reason they prefer to be described as *'people with vision impairment'* or *'people with a visual disability'*.

Who can register as blind? A person can be registrable as blind if they can only see the top letter of the eye test chart (used by opticians and doctors) or less, at a distance of three metres. This is sometimes called 3/60 vision: the patient can see at three metres distance (with fully corrected vision) what a person with normal vision can see at sixty metres. Some blind people may be able to see more than this but only if they can get the letters directly within their line of vision. In this case they are still very disabled, in that they are likely to bump into objects on either side or to trip over steps.

Who can register as partially sighted? The partial sight register is for those who, though severely visually impaired, are not so disabled as to be registrable as blind. In this instance the person may not be able to see the letters on the eye test chart nay better than *'blind'* people but the field of vision may be less limited, or they may be able to see more clearly than a blind person but only within a very limited area.

Why register? It is necessary to be registered as blind or partially sighted to qualify for most financial benefits. RNIB's Benefit Rights Unit provides detailed advice on how to claim appropriate benefits, as does the Benefits Enquiry Line, run by the Benefits Agency. Being registered means easier access to a range of special services, equipment and advice provided by social services and voluntary organisations. **Who holds the register?** Each County Council, District Council and Boroughs keep a register of blind and partially sighted people living within its area. Registration can only take place on the recommendation of an ophthalmologist, a fully qualified doctor who specialises in eyes and eye care.

Do all blind and partially sighted people register? No. Registration is voluntary and the RNIB estimates that as many as three out of four people who are eligible to register as blind or partially sighted choose not to do so. It is thought that many people are unaware of the advantages of registering and others simply do not wish to be officially *'labelled'*.

How many people are visually impaired? The OPCS Disability Survey identified over 1.7 million adults in Britain with *'a seeing difficulty'*. RNIB estimates that one million of these are registrable as blind or partially sighted. However, only a quarter of a million people are actually registered. Worldwide, there are over 42 million people who are blind, and many more who are partially sighted.

(1) Two thirds of visually impaired people are over the age of seventy five. In other words, more than one person in five over the age of seventy five is severely visually impaired. **(2)** Most of these have lost their sight in later life, as their eyes deteriorate with age. Only 9 percent of visually impaired people are born with impaired vision. **(3)** Six in every ten visually impaired people have another serious illness or disability. Many have more than one.

What are the causes of blindness? Some people are born with a visual impairment and others become visually impaired as a result of a disease or an accident. The majority of people in Great Britain lose their sight in later life and usually as part of the general ageing process. One of the most common causes of visual impairment in old age is macular degeneration, which affects the central vision.

Loss of central vision makes close-up tasks such as reading more difficult and calls for increased use of side vision. Optical aids which magnify can also be very helpful. Larger images are more likely to fall on the unaffected area and be seen. Other conditions affecting clarity of vision are cataracts, retinal detachment and diabetic retinopathy.

Two of the conditions affecting the visual field; for example which impair the peripheral or side vision, are glaucoma and retinitis pigments (tunnel vision) getting around safely is the greatest problem because the ability to see all around and capture the surrounds at a glance is reduced. Field expanders (the opposite of magnifiers) are helpful for reading. In the developing world, most blindness is caused by easily preventable malnutrition and disease. Children are as likely to become blind as adults.

What is it like to go blind? To be perfectly honest with you, there is no simply answer to this question because it is different for each individual. People who lose their sight overnight, as the result of an accident for example, tend to find things more traumatic than people who were blind or people who lose their sight gradually. They have not had as much time to adjust to their disability.

It may take them longer to realise that blindness is not the end of the world. Because of this they may be angry, frightened and depressed. Eventually, however, most people come to terms with blindness and begin to learn how to overcome the difficulties caused by their loss of sight. How easily a person overcomes those difficulties depends on a number of factors.

Each person with impaired vision is unique, with a different amount of sight, as well as different needs, abilities and expectations. Each person also receives also receives a different amount of support and help from those around them.

What are the problems caused by impaired vision? Blindness can create a number of problems but the problems will be different for each person and each person will find their own solutions. If you can not see very well you may find it difficult to find your way around a busy shopping centre safely. Even a walk to the park can be a terrifying experience.

Possible solutions include the use of a white cane (stick), a guide dog or asking a sighted friend to help. Some people find it harder to do everyday tasks like getting dressed or making a cup of tea or coffee. There are a number of solutions to these problems, including the use of special equipment, such as colour coded buttons to tell you what colour an item of clothing is.

Some people may need to learn new ways of doing something, like counting as they pour water into a teapot so that they know it is full. It takes practice and patience to master these techniques.

Reading and writing can be very difficult, all though more than 50% of people with impaired vision can actually read large print. Alternative systems include Braille and Moon, as well as audio tapes and the use of sighted readers. One of the biggest problems caused by blindness is isolation, both physical and social. It may be difficult for some blind people to get out of their own homes or to start up new friendships. The telephone and mobile may become a life-line.

Blindness can also be expensive because of the cost of special equipment, taxis, and bigger telephone bills and so on. **What can sighted people do to help visually impaired people?** The biggest compliment you can pay a visually impaired person is to forget that they have impaired vision. Do not search for substitutes for words like *'look'* and *'see'*. It is perfectly acceptable to say 'I'm as blind as a bat', even if the person you are speaking to really is!

Blind people speak the same language as everybody else. When you meet a visually person say who you are. Speak directly to that person, rather than going through a third person. Do not walk away without mentioning that you are going or they may end up talking to themselves. Always offer to help first and do not be afraid to ask: ***"How much can you see?"***

When you are guiding a blind person allow the person to take your arm. It is very frightening to be frog-marched into unknown obstacles. When you are guiding tell the visually impaired person when you are approaching steps or a kerb and say whether the steps go up or down. Mention any potential hazards and say whether they are to the left or to the right. Place a visually impaired person's hand on the back of a chair to help them orientate themselves when they sit down. Be careful not to leave doors ajar and always put things back where you found them.

If you think someone could use a bit of help. Let them tell you what kind of help they would like. This might be anything from helping with the shopping, giving them a lift to the station, mending a fuse or reading a letter.

Here are some myths about Blindness: people with disabilities often say that they can cope with the disability itself, it is the attitudes of the able-bodied which makes them *'handicapped'*. There are many misconceptions about blindness which cause problems for blind people.

Here are some of them: **Blind people see nothing** – the most common myth is that all blind people live in a world of total darkness. In fact only about 18% of blind people have no sight at all and of these most will have been blind since birth. The remaining 82% of blind people have varying degrees of vision. Some can only distinguish vague colours and shapes. More can read large print and some can even read ordinary print. The spectrum of disability is very wide. Visual impairment affects visual acuity (sharpness, clarity of vision), and visual field. Some diseases affect one of the other but it is not unusual to suffer from more than kind of eye defect.

Blind people have special gifts – this is another myth that visually impaired people are endowed with a better sense of touch, hearing, taste or even smell to compensate for the loss of vision. The reality is that, if the efficiency of one sense diminishes, people naturally make more use of their others senses to gather the information they need about the environment.

There is no _**'sixth sense'**_ to enable visually impaired people to perform such feats as getting on the right bus or preparing a meal without sight. It takes common sense and practice. Neither are visually impaired people more (or less!) clever, cheerful, outgoing, generous, creative or musical than people with sight. Blind people feel other people's faces to find out who they are.

In fact 77% of people with impaired vision actually retain enough sight to be able to recognise friends close up. If a blind person does not recognise you they will normally ask who you are. Feeling your face will not help. It is true however, that some people with multi-difficulties and disabilities may use touch as a strategy for identifying people and objects.

This is only because they are unable or unwilling to communicate through conversation. Touch may be the only way they have of understanding the world and expressing their feelings about it.

All blind people read Braille – Braille is a form of embossed script used by visually impaired people all over the world. It is thought that only 12,000+ people read Braille on a regular basis in

England. To read Braille it is necessary to develop a good sense of touch, and this is often difficult for elderly people. An easier form of embossed writing is the Moon system, which has large letters, but this is not widely used as Braille. Many visually impaired people use large print and tapes as well as, or instead of, Braille and Moon.

All blind people have their own guide dog – There are approximately 4,000 guide dog users in England, a small fraction of the total number of visually impaired people. There are several reasons for this. Most people with impaired vision are elderly, most have an additional disability or illness, and many are confined to their own homes.

They are therefore unable or unwilling to look after a healthy and lively animal which needs to be fed, and exercised regularly. Many blind people simply prefer to use other aids, such as a white cane (stick) or sighted people. For those who do have guide dogs however, they are invaluable tools and companions.

Blind people can not do normal jobs – The days when blind people were limited to careers in basket weaving and piano tuning are long gone and are no more. In this era of new technology which has brought such wonders as speaking computers there are few jobs which a visually impaired person cannot do. There are visually impaired teachers. Lawyers, social workers, computer analysts, and bankers, etcetera. However, bus or coach driving is still not advisable!!!

HEARING DISABILITY

Between 20% to 24% of people in the United Kingdom have a hearing defect (either on a temporary or permanent basis), sufficiently severe to impair normal function. 35% of those are over 55 years of age. It has been suggested that in the not too distant future this will unfortunately increase especially among the 30-35 age group due to increased levels of environmental and leisure noise.

Degrees of deafness vary greatly, as do their causes. Drug allergies, infections, physical injury, malformation of the ear and diseases like otosclerosis (which often affects elderly and middle-aged people) are all common causes of hearing problems. Many forms, however, can now be treated with drugs, surgery or hearing aids. Profoundly deaf people, especially those born deaf, may also have serious difficulty with speech. The Sympathetic Hearing Scheme has been introduced to help deaf and hard of hearing people to lead easier lives, and uses the **"ear"** symbol on stickers and cards.

Some deaf and hard of hearing people may show this card as they enter any buses – which is a very good way of telling bus drivers or a person assisting them that they have an impairment, and may need some help. So how can you best help a deaf and hard of hearing person? First of all, do not make the common mistake of shouting – it really does not help, and only causes embarrassment to those around about you.

Instead, speak slowly and clearly, but do not exaggerate your facial movements and distort your face when dealing with a deaf and hard of hearing person. Try to face the disabled person by making sure they can see your lips. Many deaf and hard of hearing people are excellent lip readers, so make sure they able to see your mouth. So you obviously need to make sure nothing else is in your mouth, or obstructing your mouth. If you are not immediately understood try to rephrase what you are saying, rather than repeating the same words in a much louder voice.

You might possibly have to write some words down, so that it is a very good idea when people carry a note pad and pen handy on them at all times, just in case they need to help a deaf and hard of hearing person at one point in their life here on earth!

Lastly, if a deaf and hard of hearing person is with one of their friends, make sure you address what you are saying to the deaf person. Their friend will still be able to follow and help out if needed. So if you happen to be a bus driver or taxi driver reading this information; it may help tremendously when you put your cab lights on to highlight your face, and the face of the deaf and hard of hearing person you are speaking and dealing with!

THE DEAF AND THE BLIND

Deaf and Blind people fall into two categories – those who have been deaf and blind from birth or from an early age, and those who have become deaf and blind in later life. The National Deaf-Blind Helpers League hold records of over 800 deaf and blind people in the United Kingdom, the great majority of who fall into the second category. Most were originally deaf or blind, and lost their other sense later in life.

There are two hand language systems used by deaf-blind people. The **"block"** system is most commonly used in communicating with people with a hearing/visual impairment and involves the writing of capital letters with the finger on the palm of the deaf-blind person's hand. So, you as a helper, when coming across a deaf-blind person if unable to speak may actually take you hand to inform you of their requirements or needs. And, you may need to reply in the same manner.

Deaf-blind customers will always need help in seating on buses and in knowing when they've reached their destination, and when to get off the bus, and yet again touch is the most important communication tool, so be extra careful how and where you touch!

MULTIPLE SCLEROSIS

Often referred to as **"MS", Multiple Sclerosis** is a progressive disease of the brain and spinal cord, caused when the tissue sheaths protecting the central nervous system become scarred and inflamed, preventing nerve impulses travelling to or from different parts of the body. It affects all ages, usually in the form of periodic attacks and remissions, sometimes made worse by stress.

It cannot be cured and may result in eventual blindness, paralysis and death, although many sufferers are only moderately affected. The symptoms are frequently mistaken for drunkenness. They include double vision, clumsiness, falling over, stiffness of limbs, trembling, giddiness, slurred speech, unusual speech patterns and difficulty in skilled or fine movement and manipulation.

The disease may also affect the frontal lobes of the brain, often producing a state of euphoria, which makes ***drunkenness*** a wrong diagnosis. Remember that people that have had too much to drink will smell of alcohol, not normally people who suffer from MS!

RHEUMATOID ARTHRITIS

Rheumatoid Arthritis affects over 1.5 million people in the UK. It is a chronic, benign, crippling illness of unknown cause which begins in the victim's 20s of 30s, and progresses without remission, though at varying rates. The disease is again, slightly more common amongst women than men. It is characterised by the symmetrical swelling of the joints in the following order – hands and fingers, elbows, shoulders, ankles and knees.

It is painful and crippling, leading to deformity and loss of function. People with an advanced condition may be confined to a wheelchair and unable to carry out the simplest tasks, like doing up buttons on a shirt or blouse and fumbling for small change.

ALZHEIMER DISEASE

In 1901 Auguste Deter, a woman in her early 50s, became the first person diagnosed with Alzheimer's disease, a form of dementia. The disease is named after the doctor who first described it, Alois Alzheimer. The disease is characterized by odd behaviour, memory problems, paranoia, disorientation, agitation, and hallucinations.

This is a degenerative disorder of the cerebral cortex in the brain which produces dementia. It usually affects old people and one of the earliest symptoms is loss of memory.

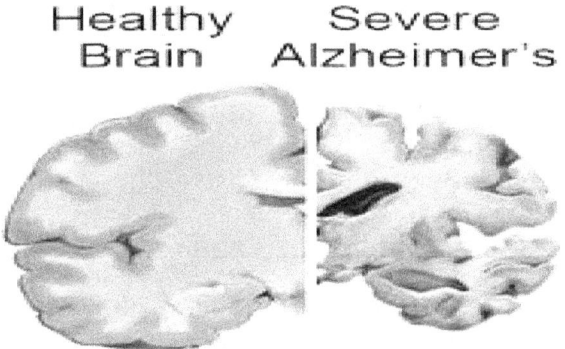

In the early stages of Alzheimer's disease, patients may experience memory impairment, lapses of judgment, and subtle changes in personality. As the disorder progresses, memory and language problems worsen and patients begin to have difficulty performing activities of daily living, such as balancing a cheque-book or remembering to take daily medications.

They may become disoriented about places and times, may suffer delusions (such as the idea that someone is stealing from them or that their spouse is being unfaithful), and may become short-tempered and hostile. During the late stages of the disease, patients begin to lose the ability to control motor functions such as swallowing, or lose bowel and bladder control.

They eventually lose the ability to recognize family members and to speak. As the disease progresses it begins to affect the person's emotions and behavior and they develop symptoms such as aggression, agitation, depression, sleeplessness, or delusions. On average, patients with Alzheimer's disease live for 8 to 10 years after they are diagnosed. However, some people live as long as 20 years. Patients with Alzheimer's disease often die of aspiration pneumonia because they lose the ability to swallow late in the course of the disease.

LEARNING DISABILITIES

Learning disabilities was previously known as mental Handicap. The World Health Organisation defines learning disabilities as *"a condition of arrested or incomplete development of the brain especially characterised by sub-normal intelligence and mental retardation"*. An IQ below 70 is often taken as an indication of learning disabilities, by which criteria there are around 1.6 million people with learning disabilities in the United Kingdom alone – 2.5% of the population.

The birth rate of people with learning disabilities has sharply declined over recent years, and now stands at 0.4% of the population. As you might expect, this means that the population is much older than previously the case. People with learning disabilities handicap may also suffer from the other problems of old age. Learning disability is a permanent condition which cannot be cured as it is a result of damage to or malfunction of the brain. In general terms, someone with learning disability would generally function at a level of ability which is significantly lower than their chronological age.

A person with a learning disability does not develop in childhood as quickly as other children and will not attain the full mental capacity of a normal adult. There are a few broad areas in which learning disability can occur:

- **Chromosomal/genetic abnormalities (one of the most common is Down's Syndrome)**
- **Infections**
- **Trauma (injury)**
- **Socio-economic factors**

People with learning disabilities can be helped to fulfill their potential and participate in general society but will be dependent upon others for assistance and support in varying degrees throughout their lives.

MENTAL ILLNESS

This is an umbrella term for a number of conditions which can have extremely complex causes, and range from severe depression to schizophrenia and dementia. For example, 5% of all people over 65 have mild dementia. A further 5% suffer severe dementia – perhaps over 1.3 million in total.

Symptoms will vary according to the form of mental illness experienced, but may include irrationality, anxiety, delusions, depression, disorientation, hallucination and bizarre speech and conduct. Some of the most common types of mental illness include anxiety, depressive, behavioral, and substance-abuse disorders.

There is no single cause for mental illness. Rather, it is the result of a complex group of genetic, psychological, and environmental factors. While everyone experiences sadness, anxiety, irritability, and moodiness at times, moods, thoughts, behaviours, or use of substances that interfere with a person's ability to function well physically, socially, at work, school, or home are characteristics of mental illness.

There is no one test that definitively indicates whether someone has a mental illness. Therefore, health-care practitioners diagnose a mental disorder by gathering comprehensive medical, family, and mental-health information. Talk therapy (psychotherapy) is usually considered the first line of care in helping a person with a mental illness. It is an important part of helping individuals with a mental disorder to achieve the highest level of functioning possible.

Psychotherapies that have been found to be effective in treating many mental disorders include family focused therapy, psycho-education, cognitive therapy, interpersonal therapy, and social rhythm therapy. Medications may play an important role in the treatment of a mental illness, particularly when the symptoms are severe or do not adequately respond to psychotherapy.

A variety of factors can contribute to the prevention of mental-health disorders. Individuals with mental illness are at risk for a variety of challenges, but these risks can be greatly reduced with treatment, particularly when it is timely. Mental illness refers to all of the diagnosable mental disorders. Mental disorders are characterized by abnormalities in thinking, feelings, or behaviours.

SCHIZOPHRENIA

This is the most frequent form of psychotic disease or *"madness"*. Many people experience the onset of schizophrenia in their late teens or early twenties, and the symptoms include inappropriate responses in thinking, speech and behaviour, sudden and unprovoked mood changes and irrational conduct. 25% of schizophrenics are considered to be severely disabled and sufferers are often vulnerable and alienated.

For reasons which are still not entirely understood, there is sometimes an easing of the disease in middle age – you hear such people referred to as *"burnt-out"* schizophrenics. All people suffering from mental illness should be treated with courtesy and care, as many will be highly sensitive to any perceived hostility. Schizophrenia affects about 1.1% of the world's population. Schizophrenia is most commonly diagnosed between the ages of 16 to 25.

Schizophrenia can be hereditary (runs in families), and it affects men 1.5 times more commonly than women. Schizophrenia and its treatment have had an enormous effect on the world's economy, costing billions each year. There are five types of schizophrenia (listed below). They are categorised by the types of symptoms the person exhibits when they are assessed:

- **Paranoid schizophrenia**
- **Disorganized schizophrenia**
- **Catatonic schizophrenia**
- **Undifferentiated schizophrenia**
- **Residual schizophrenia**

Paranoid-type schizophrenia is distinguished by paranoid behaviour, including delusions and auditory hallucinations. Paranoid behaviour is exhibited by feelings of persecution, of being watched, or sometimes this behaviour is associated with a famous or noteworthy person a celebrity or politician, or an entity such as a corporation. People with paranoid-type schizophrenia may display anger, anxiety, and hostility. The person usually has relatively normal intellectual functioning and expression of affect.

A person with disorganized-type schizophrenia will exhibit behaviours that are disorganized or speech that may be bizarre or difficult to understand. They may display inappropriate emotions or reactions that do not relate to the situation at-hand. Daily activities such as hygiene, eating, and working may be disrupted or neglected by their disorganized thought patterns.

Disturbances of movement mark catatonic-type schizophrenia. People with this type of schizophrenia may vary between extremes: they may remain immobile or may move all over the place. They may say nothing for hours, or they may repeat everything you say or do. These behaviours put these people with catatonic-type schizophrenia at high risk because they are often unable to take care of themselves or complete daily activities.

Undifferentiated-type schizophrenia is a classification used when a person exhibits behaviours which fit into two or more of the other types of schizophrenia, including symptoms such as delusions, hallucinations, disorganized speech or behaviour, catatonic behaviour.

This is when a person has a past history of at least one episode of schizophrenia, but the currently has no symptoms (delusions, hallucinations, disorganized speech or behaviour) they are considered to have residual-type schizophrenia. The person may be in complete remission, or may at some point resume symptoms.

STROKES

Strokes are the sudden manifestation of vascular disease in the brain, either a cerebral heamorrhage (bleeding into the brain) or cerebral thrombosis (a blood clot in a cerebral artery) Strokes vary vastly in effect. Some are fatal; others leave the person with varying degrees of paralysis, whilst other people make a full recovery.

There are three main physical symptoms – paralysis along one side of the body or face, loss of, or slurring of speech and incontinence. In some cases there may be some degrees of mental impairment. These symptoms can be easily mistaken for drunkenness, but close observation should soon reveal the truth.

What is a Heart attack and how does it happen?

A Heart attack, or myocardial infarction, is probably the number one killer of both men and women throughout the world today. Each year, many people suffer heart attacks and many of these are fatal. Most of the deaths from heart attacks are caused by ventricular fibrillation of the heart that occurs before the victim of the heart attack can reach an emergency room. Those who reach the emergency room have an excellent prognosis; survival from a heart attack with modern treatment should exceed 90%.

The 1% to 10% of heart attack victims who die later includes those victims who suffer major damage to the heart muscle initially or who suffer additional damage at a later time. Fortunately, procedures such as coronary angiogram and PTCA (coronary balloon angioplasty), and clot dissolving drugs are available that can quickly open blocked arteries in order to restore circulation to the heart and limit heart muscle damage. In order to optimally benefit heart attack victims and limit the extent of heart damage, these treatments to open blocked arteries should be given early during a heart attack.

Blood pressure is not a reliable measurement of whether one is having a heart attack. Blood pressure during a heart attack can be low, normal, or elevated. Knowing the early warning signs of heart attack is critical for prompt recognition and treatment. Many heart attacks start slowly, unlike the dramatic portrayal often seen on television or movies.

A person experiencing a heart attack may not even be sure of what is happening. Heart attack symptoms vary among individuals, and even a person who has had a previous heart attack may have different symptoms in a subsequent heart attack. Although **chest pain** or pressure is the most common symptom of a heart attack, heart attack victims may experience a diversity of symptoms that can include:

Chest discomfort, manifest as pain, fullness, and/or squeezing sensation of the chest; jaw pain, toothache, headache; shortness of breath; nausea, vomiting, general epigastria (upper middle abdomen) discomfort; sweating; heartburn and/or indigestion; arm pain (more commonly the left arm, but may be either arm); upper back pain; general malaise (vague feeling of illness); and no symptoms (approximately one quarter of all heart attacks are silent, without chest pain or new symptoms and silent heart attacks are especially common among patients with diabetes mellitus).

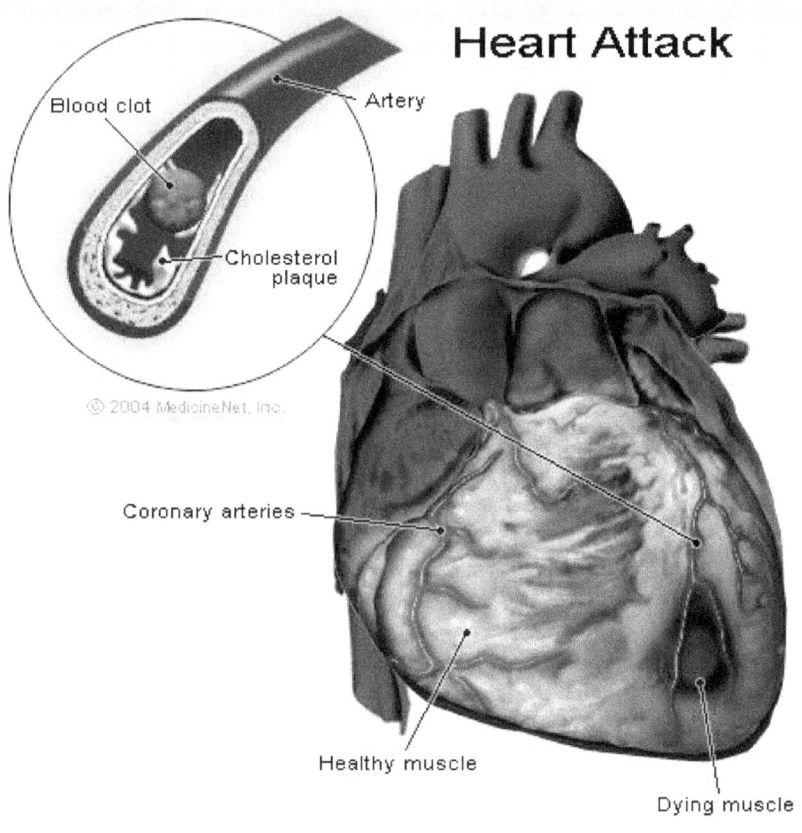

Even though the symptoms of a heart attack at times can be vague and mild, it is important to remember that heart attacks producing no symptoms or only mild symptoms can be just as serious and life-threatening as heart attacks that cause severe chest pain. Too often patients attribute heart attack symptoms to **"<u>anxiety</u>,"** "indigestion," **"<u>fatigue</u>,"** or **"<u>stress</u>,"** and consequently delay seeking prompt medical attention. ***One cannot overemphasize the importance of seeking prompt medical attention in the presence of symptoms that suggest a heart attack.***

Early diagnosis and treatment saves lives, and delays in reaching medical assistance can be fatal. A delay in treatment can lead to permanently reduced function of the heart due to more extensive damage to the heart muscle. Death also may occur as a result of the sudden onset of arrhythmias such as ventricular fibrillation.

What should you do if you experience these symptoms? Medical Doctors agree that if you're in doubt, get checked out anyway. Even if you're not sure if something is really wrong, you should call **<u>999</u>** or **<u>911</u>** if you experience heart attack symptoms. Prompt administration of drugs can help restore circulation to the heart and increase your chances of survival.

PARKINSON'S DISEASE

This is a chronic, disabling brain disease affecting elderly, middle-aged and young people, in descending order of frequency. Most cases are due to degenerative vascular disease in the brain, or trembling limbs. Some sufferers are bent over, with a characteristic shuffling gait and have difficulty in controlled movement.

Muscular rigidity affects facial muscles, giving victims a flat, glazed expression, which may be characterised by the slowness with which smiles or frowns form and their long persistence. This is often confused with signs of mental inadequacy. It should be stressed that there are no mental effects.

EPILEPSY

What is Epilepsy? Epilepsy is the name given to recurrent *'seizures'* (the modern term for 'fit') which are known to have started in the brain. **EPILEPSY IS A DISORDER OR CONDITION AND NOT A DISEASE!** Epilepsy is not a disabling condition, nor is it a single disease and exact figures are hard to establish or estimate. Some people may face restrictions on driving, and will be more than usually dependent on other modes of transport.

Epilepsy is essentially a condition of disorder consciousness. It manifests in a variety of ways – from the sufferer who misses the odd word in a sentence to the person who has a violent fit. Epilepsy certainly does not result in any mental or intellectual impairment. Indeed, there is the argument that the exact opposite is true, as many exceptional distinguished people have been and are epileptic.

People like Martin Luther of the Reformation; Socrates; Napoleon Bonaparte; Sir Isaac Newton; Agatha Christie; Charles Dickens; Alexander the Great; Neil Young; Danny Glover; Leonardo Da Vinci; Michelangelo; Vincent van Gogh, Alfred Nobel; Julius Caesar; Edgar Allan Poe; Theodore Roosevelt; Aristotle; Bud Abbott; Lewis Carrol; Richard Burton; Hannibal; Lord Bryon; Nicolo Paganini; Peter Tchaikovsky; Sir Walter Scott; Robert Schumann; Margaux Hemingway; Charles the V of Spain; George Frederick Handel; Hugo Weaving and so many others.

Epileptic seizures take three forms. The first and most serious is called a ***"grand mal"*** - which begins with a sudden, short-lived, often euphoric sense of detachment, followed by rigidity and falling over, then uncontrolled spasms, followed by sleep or coming around in a comparatively short time. The second, known as ***"petit mal"*** involves a sudden loss of consciousness, without falling to the ground, which often happens in the middle of a conversation or action.

The person may drop their bags to the ground or stare blankly into space. The third form is known as a *"focal seizure"* and is almost the result of organic brain disease. There may be involuntary spasms of that part of the body which is controlled by the damaged brain section, without loss of consciousness.

Epilepsy responds well to pre-emptive medication. In the event of a person experiencing a fit, you should act to prevent the person injuring themselves, by moving away any obstacles. **How is a diagnosis made when someone is suffering from epilepsy?** Usually this is done and (most importantly) through Medical or Family History; Observation of Seizure(s); Physical Examination; Blood Tests; EEG; **(and when relevant)** CT Scan; MRI Scan and by Ambulatory Monitoring.

This diagnosis is a clinical decision. Some causes of epilepsy are Head Injury; Brain Infection – e.g. Meningitis; Stroke; Brain Damage; Drugs and Alcohol; Bio-Chemical Imbalance; Hormonal Changes; Other Conditions – e.g. Cerebral Palsy, Tuberous Sclerosis. * **Over 50% of cases – no known cause** *

Some causes of individual seizures ('triggers') are Forgotten or Incorrect Medication; Lack of Sleep; Stress; Excitement; Boredom; Alcohol; Flashing Lights – only 3-5% of people with epilepsy are photosensitive; Drugs; Diet – Lack of Food and Illness.

PARAPLEGIA

Paraplegia means paralysis of the body from the waist down. It occurs as a result of diseases and injuries which affect the spinal cord. People with paraplegia are confined to wheel-chairs and cannot use their legs at all – which is by no means true of all wheel-chair users.

OLD AGE

At present, some 20% of the population is over 65 years of age and the percentage continues to grow. As we have shown, many elderly people will suffer a variety of disabling conditions and require extra consideration. As we age, bone density decreases as well and can lead to osteoporosis, a condition in which the bones become fragile and weak, and are more prone to fractures.

Many people throughout the world have or are at risk for osteoporosis, and it is more common in women than in men. Exercise can increase bone strength and density. Weight-bearing activity in particular is useful as these causes the bones to work harder. Exercise is an important key to aging successfully. It's never too late to start. In the following slides we will look at how our bodies age, the benefits of exercising into old age, and tips to get started on your fitness journey. Strength training as well strengthens muscles and helps strengthen bones. As we age, balance decreases and falls can lead to fractures.

As we age muscle mass decreases. Between the third and eighth decades of life, we lose up to 15% of our lean muscle mass, which contributes to a lower metabolic rate, as we get older. Maintaining muscle strength and mass helps burn calories to maintain a healthy weight, strengthens bones, and restores balance. It's never too late to exercise and build muscle.

The body is responsive to strength training at any age. Strength training can help reduce symptoms of some common problems we encounter as we age including arthritis, diabetes, osteoporosis, obesity, back pain, and depression. Strength doesn't just involve building large muscles. Lifting weights just two or three times a week can increase strength by building lean muscle. Studies have shown that even this small an amount of strength training can increase bone density, overall strength, and balance.

It can also reduce the risk of falls that can lead to fractures. Just as muscle mass declines with age, so does endurance. The good news is that the body also responds to endurance fitness training such as walking. Any activity that increases heart rate and breathing for an extended period is considered endurance exercise. In addition to walking, swimming, cycling, dancing, and tennis are all endurance activities.

World health statistics estimate more than one-third of people over the age of 65 falls each year, often resulting in injuries such as hip fractures which are a major cause of surgeries and disability among the elderly. Balance and strength exercises can help maintain balance and reduce the risk of falling. Along with muscle mass and endurance, flexibility also decreases as we age. But like strength and endurance, flexibility too can be improved. Increased flexibility allows for more freedom of movement and greater range of motion. Areas to pay attention to are the neck, shoulders, elbows, wrists, hips, knees, and ankles.

Exercise helps with cognitive function. Studies have shown that regular physical activity can slow declines in memory and protect against dementia. Exercise has been shown to improve mood. Depression is common in older adults, and exercise can have an antidepressant effect. It is thought that exercise may increase serotonin in the brain, which leads to better moods and less depression.

Before starting any exercise program, talk to your doctor to find out what activities is right for you. It's important to start slowly, and build gradually. Doing, too much exercise too soon can result in injury. Even a five-to-ten minute walk is a good starting place, and you can build from there. Motivate yourself with goals.

Any activity that increases heart rate and breathing for an extended period is considered endurance exercise. Endurance and aerobic exercises are good for your heart, lungs, and the circulatory system. Endurance gives you stamina for daily tasks, and can prevent many aging-related diseases such as diabetes, heart disease, and strokes.

Walking, running, cycling, swimming, aerobics classes, and tennis are all types of endurance exercise. Many gyms and senior centres offer exercise classes for seniors. Endurance exercise does not have to be strenuous to be beneficial. No matter what your age, exercise is good for you. It's never too late to start, and you can benefit from strength and resistance training, stretching and flexibility exercise, and endurance and aerobic exercises. Find exercise that you enjoy that fits into your schedule and get started, and Go for It!!!

MUSCULAR DYSTROPHY

In many cases this seems to be a hereditary disease which causes the wasting of certain muscles – typically of the hips, thighs and calves of young children, and the arms, spine, shoulders and face of older people.

Importantly, the disease does not cause any mental or intellectual deterioration, but patients can become physically helpless, unable to perform the smallest personal tasks and confined to a wheelchair.

In the early stages, the symptoms can be similar to those of Multiple Sclerosis – clumsiness and falling over being typical. Most victims are likely to be children. Duchene muscular dystrophy is the most common kind of muscular dystrophy affecting children. Myotonic dystrophy is the most common of these diseases in adults.

There is no specific treatment for any of the forms of muscular dystrophy. Physical therapy to prevent contractures (a condition in which shortened muscles around joints cause abnormal and sometimes painful positioning of the joints) orthoses (orthopedic appliances used for support) and corrective orthopedic surgery may be needed to improve the quality of life in some cases.

The cardiac problems that occur with Emery-Dreifuss muscular dystrophy and Myotonic dystrophy may require a pacemaker. The myotonia (delayed relaxation of a muscle after a strong contraction) occurring in Myotonic dystrophy may be treated with medications such as phenytoin or quinine. The prognosis (outlook) with muscular dystrophy varies according to the type of muscular dystrophy and the progression of the disorder.

Some cases may be mild and very slowly progressive with normal lifespan, while other cases may have more marked progression of muscle weakness, functional disability and loss of ambulation. Life expectancy depends on the degree of progression and late respiratory deficit. In Duchene muscular dystrophy, death usually occurs in the late teens to early 20s. Muscular dystrophy is abbreviated as MD.

OSTEOARTHRITIS

This is a condition caused by the degenerative inflammation of the joints. Classic symptoms are pain and stiffness in the legs, hips, hands and shoulders, with a concurrent loss of fine movements, making step climbing, gripping and rooting in bags or purses difficult.

The causes are unknown. Between 80-90% of all people over the age of 60 will suffer from this condition, and it has been estimated that over 10 million people in England will suffer from varying degrees of impairment caused by osteoarthritis. This condition is more common in women than in men – all though it is not exclusively a disease of the elderly.

CALLING FOR AN AMBULANCE

It is not usually necessary to call for an ambulance when managing epileptic seizures. However, it should be considered if a doctor cannot attend straight away in the following circumstances:-

(1) It is the first seizure, the cause of which is uncertain and needs investigation. (2) Injuries have occurred during the seizure, for example: a cut that needs stitching. The elderly are especially at risk from falls. (3) The convulsive part of the seizure shows no sign of stopping after 5-6 minutes or is 2 minutes longer than is usual for that person. If a second seizure occurs without the person regaining consciousness. (5) If any doubt at all – then please call for an ambulance.

ABOUT THE AUTHOR

Daniele Luciano Moskal was born in Supino, Italy, and holds a B.A degree in Writing & Publishing, Media Studies and Religious Studies. He also attended Oak Hill Theological/Bible College in London, England. He is the prolific author of many non-fiction books and many Children's books, and the 'ghost writer' of many other titles all in print.

Information at:

http://www.dlm-evangelical-ministries.org/

THANKS FROM THE AUTHOR

If you enjoyed this book, and found it useful or otherwise, then I'd really appreciate it if you would post a short review on Amazon or Lulu. I do read all the reviews personally so that I can continually write what people are wanting. God bless and thanks for your support!

www.ingramcontent.com/pod-product-compliance
Lightning Source LLC
Chambersburg PA
CBHW070431180526
45158CB00017B/969